Coding for Kids in C#

Table of Contents

Disclaimer

Copyright © 2022
All Rights Reserved

Free Gift

We do want you to succeed in coding. To ensure your success, we are giving you a free list of projects that you can work on once you are completed with this book.

https://coding.gr8.com/

Introduction

Programming was a relatively unknown term in the past, as computer programmers were thought of as hardcore geeks who knew the ins and outs of the machine right in front of your eyes every day.

Nowadays, everything has shifted. In fact, Programming is one of the top skills any person can have, and starting out as a teenager is a massive plus, since you have the biggest advantage on your side - time.

Not only does programming help you with practical & everyday situations, it could also help you land the job of your dreams. Most programmers throughout the world have top-paying jobs with excellent benefits - like remote work, flexible hours, and amazing projects.

In this book, we will teach you programming in one of the most popular, practical, and effective programming languages - C# (read as C-sharp). With it, you will learn all of the basics of programming, but also dabble with slightly more difficult topics.

SET UP WORKSPACE

What You Need

- A Laptop/Desktop Computer
- Internet Connection
- Web browser installed (Firefox, Chrome or Internet Explorer)

Download and Install Visual Studio

Download the latest version of an IDE of your choice. I recommend choosing between one of the following:

- Visual Studio
- Tabnine
- SlickEdit
- Elipse aCute

I personally recommend using Visual Studio. You can download the latest version below:

https://visualstudio.microsoft.com/downloads/

How to Start Using Visual Studio

Open Visual Studio from your installation folder or Start Menu.

Choose "Create a new project"

Choose C# and select "Console Application".

Name your project and set a directory where you want the project to be placed.

Then, your new project will open in a new window

```
gram.cs  ⊦ ✕
ConsoleApp2
1    ⊟using System;
2     using System.Collections.Generic;
3     using System.Linq;
4     using System.Text;
5     using System.Threading.Tasks;
6
7    ⊟namespace ConsoleApp2
8     {
         0 references
9    ⊟     class Program
10           {
             0 references
11   ⊟         static void Main(string[] args)
12               {
13               }
14           }
15    }
16
```

Write your first lines of code, for example:

- Console.WriteLine("Hello World");

```
ng System;
ng System.Collections.Generic;
ng System.Linq;
ng System.Text;
ng System.Threading.Tasks;

espace ConsoleApp2

0 references
class Program
{
    0 references
    static void Main(string[] args)
    {
        Console.WriteLine("Hello World!");
    }
}
```

Press the "Start" button in the top menu, and your code will start running.

Intro to C#

C# (pronounced as C-Sharp) is an object-oriented programming language that runs on the .NET Framework. It was made by Microsoft in 2002, and since then, it has had 8 versions.

Where is C# Used?

C# is a general-purpose language, meaning that it has a lot of use cases. It can be used for making:

- Mobile apps
- Desktop apps
- Web apps
- Games
- Virtual reality apps
- Database apps
- Augmented reality simulations
- And much more.

Why Should You Use C#?

C# is one of the most popular programming languages in the world. According to most statistics it's easily in the top 10. Moreover, the language is not difficult to learn, and has massive community support on multiple forums and websites. This means that there's always a channel to contact in case you have an issue.

The language is also very similar to other programming languages, like C, C++, Java, and Kotlin, so you'll have no trouble learning another one of them once you get the hang of C#.

Example Code in C#

```csharp
using System;

namespace ConsoleApp2
{
    class Program
    {
        static void Main(string[] args)
        {
            Console.WriteLine("Hello World!");
        }
    }
}
```

- Code output:

```
Hello World!
```

Explaining individual components:

using System;

- We're taking all the classes and functions from the **System** class.

namespace ConsoleApp2;

- Namespace refers to the name of the project - it's simply the label of the application (i.e. how you named it).

class Program

- A container for all the methods and the code you'll write. Everything you write in C# must be within a class. We'll talk more about classes in a later chapter.

Simple C# Code Example

Now that we've gotten the basics out of the way and explained what each line of code does, let's take a look at a **slightly** more complicated C# code example:

```csharp
using System;

namespace ConsoleApp2
{
    class Program
    {
        static void Main(string[] args)
        {
            var a = 5 + 5;
            Console.WriteLine(a);
        }
    }
}
```

So, what will be the example output?

- If you answered 10, your answer is correct

This was a really simple arithmetic exercise. We declared a **variable** called **a,** and we assigned a **value** of **5+5** to it (which is 10).

- Code output:

```
10
```

Conditionals

If you've learned the basics of mathematical logical conditions, conditionals in C# will probably be incredibly easy for you. Here is an example from classic mathematics:

- **a > b:** a is bigger than b
- **a <= b:** a is bigger than or equal to b
- **a == b:** a is equal to b
- **a != b:** a is not equal to b

Based on these conditions (i.e. whether they are fulfilled or not) you can make your code take a totally different direction. With the help of conditions and **conditional statements,** you can make the code decide for itself and do a specific thing. These are the conditional statements in C#:

- **if(condition):** if the condition inside the parentheses - the symbols "(" and ")" - is true, then the code below the statement will execute
- **else:** the else condition can **only** be written after an **if** condition. If the condition in the **if** statement is not true, then the **else** code block will get executed

- **else if(condition):** you can line up an **else if** only after an **if** and before an **else.** If the first "if" statement doesn't fulfill the condition, then the condition for the **else if** block will get checked and the code will be executed.
- **switch:** the **switch** conditional statement is used when you have many different cases & a variety of conditions that could be met. You can totally avoid using the **switch** case, though there are specific scenarios where it's quite useful.

Examples of Conditionals

Now that we've got the theory out of the way, let's get a more practical sense of how conditionals in C# work:

```csharp
int a = 50;
int b = 49;

if(a > b) {
    Console.WriteLine("a is greater than b");
}

else {
    Console.WriteLine("b is greater than a");
}
```

In the example above, we have two variables - **a** and **b** - and one of them is larger (50 is larger than 49). Thus, the code after the **if** statement - the **Console.WriteLine()** - will get executed. Then, it would print out "a is larger than b".

If **b** was larger than **a** (i.e. if **b** was at least 51), then the code in the **else** statement would get executed. It would print out "b is larger than a"

- Code output:

```
a is greater than b
```

Now that you've got a basic idea of how an **if** and **else** statement work, let's take a look at a slightly more complex example:

```
int a = 20;
int b = 5;

if (a > b) {
    Console.WriteLine("a is bigger than b");
}

else if(a == b) {
    Console.WriteLine("the numbers are equal");
}
else {
    Console.WriteLine("b is bigger than a");
}
```

A beginner programmer wouldn't see the need for the **else if** block in this scenario. In the first **if** statement, you're checking if **a** is bigger than **b** - if **a** is NOT bigger than **b**, then it has to be smaller, right?

Absolutely not! **b** could also be equal to **a,** which is why there is an **else if** code block. Even though this is an incredibly simple (and frankly bizarre) example with only 3 possible outcomes, coding doesn't simply narrow down to **true** and **false**. There are many examples where another conditional statement could make the project at hand much easier.

- Code output:

```
a is bigger than b
```

Example 2: A simple program to check one's age and tells he/she is an adult or not. Example code is given below.

```csharp
using System;

namespace MyApplication
{
  class Program
  {
    static void Main(string[] args)
    {
      int age = 15;
      if (age > 18)
      {
        Console.WriteLine("Person is an adult");
      }
      else
      {
        Console.WriteLine("Person is a child");
      }
    }
  }
}
```

In this example code variable **age** is assigned with a value of 15 and under the if else condition it checks whether the age variable is more than 18. Since age is more than 18 and the condition becomes true, the codes inside the if block gets executed. The "Person is an adult" statement gets printed. If anyhow the variable age becomes lower than 18 and the condition becomes false, the code under else block get executed and statement "Person is a child" gets printed.

Output is as follows:

```
Person is a child
```

Example 3: A simple program to check whether one has passed the exam by scoring more than 35 marks from the examination paper.

Sample Code is as follows:

```csharp
using System;

namespace MyApplication
{
    class Program
    {
        static void Main(string[] args)
        {
            int marks = 50;
            if (marks > 35)
            {
                Console.WriteLine("Congratulations. You've passed the exam.!!!");
            }
            else
            {
                Console.WriteLine("Appreciate your efforts. Please try again.");
            }
        }
    }
}
```

In this example code, variable **marks** is assigned with a value of 50 and under the if else condition it checks whether the **marks** variable is more than 35 to check whether the person have passed the exam. Since marks is more than 35 and the condition becomes true, the codes inside the if block gets executed. Then

the statement "Congratulations. You've passed the exam.!!!"
gets printed.

Output is as follows:

```
Congratulations. You've passed the exam.!!!
```

Example 3: A simple program to initial 2 variables with values of
10, 20 respectively and add 3 to the second value if the sum of 2
numbers are more than 25.

Sample code is as follows:

```
using System;

namespace MyApplication
{
  class Program
  {
    static void Main(string[] args)
    {
      int x = 10;
      int y = 20;
      int total = x+y;
      if (total > 25)
      {
        y=y+3;
        Console.WriteLine("Second Number is " + y);
      }
      else
      {
        Console.WriteLine("Second Number is " + y);
      }
    }
  }
}
```

Here the two variables **x** and **y** are initiated as 10 and 20
respectively. Another variable **total** is assigned with the sum of **x**

and **y**. Under the If condition it checks whether the total is greater than 25 or not. Since here sum of x and y variables are 30, the total is higher than 25. So the if condition becomes true and the codes inside the if condition gets executed. There y value gets incremented by 3 and the final value gets printed as 23. Essentially 3 is added to the number if the sum is greater than 30.

Output is as follows:

```
Second Number is 23
```

Switch Examples

The example code shown below takes the number of weekdays to calculate the weekday name:

```
namespace MyApplication
{
    class Program
    {
        static void Main(string[] args)
        {
            int day = 5;
            switch (day)
            {
                case 1:
                    Console.WriteLine("Monday");
                    break;
                case 2:
                    Console.WriteLine("Tuesday");
                    break;
                case 3:
                    Console.WriteLine("Wednesday");
                    break;
                case 4:
                    Console.WriteLine("Thursday");
                    break;
                case 5:
                    Console.WriteLine("Friday");
                    break;
                case 6:
                    Console.WriteLine("Saturday");
                    break;
                case 7:
                    Console.WriteLine("Sunday");
                    break;
            }
        }
    }
}
```

In this example, integer type variable day is assigned with the value of 5. Under the switch block, different cases are listed out. Case value is equal to the value of the day, which is checked under the switch block gets executed. Since there is a break command at the end of case 5 other cases will not be checked after case 5 and will terminate the program in running under the switch block commands.

This executes the block of code under case 5 and prints "Friday".

The output is as follows:

```
Friday
```

Else if condition Examples

A simple program to get a grade for the examination marks as per the following conditions.

If 0 < marks <= 35, it prints "Lower Pass"

    35 < marks <=60, it prints "Good Pass"

    60 < marks <= 100, it prints "Excellent Pass"

Sample code is as follows:

```
using System;

namespace MyApplication
{
  class Program
  {
    static void Main(string[] args)
    {
      int marks = 50;
      if (marks > 0 && marks <= 35)
      {
        Console.WriteLine("Lower Pass");
      }
      else if (marks > 35 && marks <= 60)
      {
        Console.WriteLine("Good Pass");
      }
      else if (marks > 60 && marks <= 100)
      {
        Console.WriteLine("Excellent Pass");
      }
      else
      {
        Console.WriteLine("Marks are not valid");
      }
    }
  }
}
```

Here the variable **marks** are assigned with the value 50. There are different conditions to check whether to which range the marks variable belongs to. In the first condition it checks whether the marks is greater than 0 and less than or equal to 35. If that condition satisfies the user has obtained a "lower pass". Similarly, the passes are given to the user after checking with the ranges of marks. Since the **marks** variable has a value of 50, it satisfies the condition of marks greater than 35 and less than or equal to 60. Hence, the final output becomes 'Good Pass'. Output is as follows:

```
Good Pass
```

Examples to Practice

1. Write down a long string (preferably, find a quote on the internet). Then, use an if/else code block to check if the number is odd or even. If it's even, print "even", otherwise, print "odd" (hint: use **if (a % 2 == 0)** to check if a number is even).

2. Create a simple C# and assign values to 4 variables (a, b, c, d). Check if a is bigger than b, or if c is bigger than d. If both fail, print out "test failed".

Nested Conditionals

You've now seen the importance of conditionals. If you're frequently around technology, you're probably going through conditionals every day - when opening websites, logging in, registering, playing your favorite video game, and much more.

However, conditionals alone won't do the job in many cases, which is why we use **nested conditionals.** These are the same conditionals we learned about above, but they're **nested** in each other. Let's take a look at a simple example.

```
int a = 20;
int b = 10;
int c = 25;

if(a > b) {
    if(a > c) { // nested conditional
        Console.WriteLine("a is bigger than b and c");
    }
    else { // another nested conditional
```

```
        Console.WriteLine("a is bigger than b, but
not bigger than c");
        }
}
```

With the first **if** block, we checked if **a** is bigger than **b**. Then, we did a second check to see the size of **c**, and in this case (**a** is smaller than **c**), we will get the second **WriteLine()** as output →

"a is bigger than b, but not bigger than c".

- Code output:

```
a is bigger than b, but not bigger than c
```

Nested Conditionals Examples

- A simple program checking the length of names.

```
string name1 = "Marco";
string name2 = "Johnny";
string name3 = "Sebastian";

if(name3.Length > name2.Length) {
      if(name3.Length > name1.Length) {
            Console.WriteLine("Sebastian is the
longest name");
      else if(name3.Length < name1.Length) {
            Console.WriteLine("Marco is longer than
Sebastian & smaller than Johnny");
}

else {
      Console.WriteLine("Sebastian's name is not
longer than Johnny's name");
}
```

- Code output:

```
Sebastian is the longest name
```

- A slightly more advanced example - checking the values of numbers with nested conditionals.

```csharp
int number = 12;

if (number < 5)
    {
    Console.WriteLine("{0} is less than 5",
number);
    }
else
    {
    Console.WriteLine("{0} is greater than or
equal to 5", number);
}

Console.WriteLine("The program ends here!");
```

- Code output:

```
12 is greater than or equal to 5
The program ends here!
```

Examples to Practice

1. Using nested conditional, make a program that compares the values of 3 variables. With nested if, else if, and else loops, find the highest, the lowest, and the value in between.

2. Make a string long 10 - 20 characters (think of some sentence). If the string has an even number of characters (hint: use **if(a % 2 == 0)** to check if a number is even), then nest another conditional to check if the length is equal to 14, 16, or 18. Print out the adequate result.

A Short message from the Author:

Hey, are you enjoying the book? I'd love to hear your thoughts!

Many readers do not know how hard reviews are to come by, and how much they help an author.

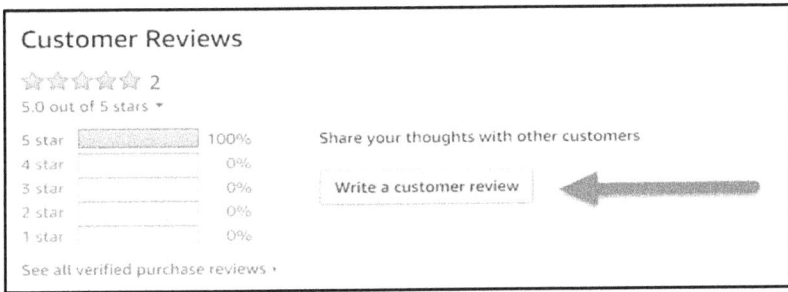

Customer Reviews

★★★★★ 2
5.0 out of 5 stars ▾

5 star	100%
4 star	0%
3 star	0%
2 star	0%
1 star	0%

See all verified purchase reviews ›

Share your thoughts with other customers

Write a customer review

I would be incredibly thankful if you could take just 60 seconds to write a brief review on Amazon, even if it's just a few sentences!

>> Click here to leave a quick review

Thank you for taking the time to share your thoughts!

Your review will genuinely make a difference for me and help gain exposure for my work.

Loops

Loops in C# are used when you want to execute some written code multiple times. Programming offers us the power of automation, and loops are the perfect example of how automation makes everything easier. First, we'll take a look at the different types of loop, and how they work, and then we'll look at a practical example.

- **while(condition):** these types of loops (often called **while** loops) will continue executing the code inside of them until the condition is false.
- **for(statement 1, condition, statement 2):** known as **for** loops, these are more popular loops as they have more use cases. They work like this:
    - **statement 1:** a statement that executes before the loop starts
    - **condition:** the loop will continue running the code until the condition is false.
    - **statement 2:** a statement that executes when the loop ends an iteration.

Here is an example of a while loop:

```
int i = 1;
while( i < 5)  { // smaller than or equal to 5
        i++; // the same as i = i + 1, increases i by
1
}
Console.WriteLine(i); // what is the value of i
now?
```

After the loop ends, i is equal to 5. Here's the output:

5

A **for** loop looks slightly more complicated, but it's used in a wider variety of cases:

```
for(int i = 1; i <= 5; i++) {
        Console.WriteLine(i); // will print 1, 2, 3,
4, 5
}
```

With the **for** loop, we declared the value of i in the first statement (see above), we defined the condition - i is smaller than 5 - and we set the second statement - increase i by one.

If you could define the code above in one simple sentence, it would be:

- Make the number i equal to 1, and increase it until it reaches 5.

As you can already imagine, there are many use cases for loops. We've taken a simple real-world example of code that shows the importance of loops (a simple countdown program).

```
Console.WriteLine("Counting down from 5 to 1: ");
Console.WriteLine("5");
Sleep(1000); // this function tells the program to
make a 1 second pause
Console.WriteLine("4");
Sleep(1000);
Console.WriteLine("3");
Sleep(1000);
Console.WriteLine("2");
Sleep(1000);
Console.WriteLine("1");
Console.WriteLine("Countdown successful!");
```

A 5 second countdown is hardly a difficult task, but in this case, it required more than 10 lines of code - for a simple countdown! Which is exactly why we need loops.

```
Console.WriteLine("Counting down from 5 to 1: ");
for(int i = 5; i >= 1; i = i - 1) {
    Console.WriteLine(i);
    Thread.Sleep(1000); }
Console.WriteLine("Countdown successful!");
```

We've now successfully cut the code in half (from 11 lines to 5), and made it significantly easier to read.

- Code output

```
Counting down from 5 to 1:
5
4
3
2
1
Countdown successful!
```

Examples For Loops

There are many situations where loops will be useful.

- **Program that calculates anything iteratively.**

Let's say that you want to multiply every second number from 1 to 10 with 5. Without loops, it would be quite a tedious task, but with them, it would look something like this:

```
int finalValue = 0;

for(int i = 1; i <= 10; i = i + 2) {
    finalValue = finalValue + (i * 5);
}
Console.WriteLine(finalValue);
```

- Code output:

```
125
```

There's no need for loops to be used strictly automatically. You can also use them with strings. Let's say you want to print every third letter from some text.

- **C# program that takes every third letter with loops**

```
string text = "Some random text";

for(int i = 0; i < text.Length; i = i +3) {
    Console.WriteLine(text[i]);
}
```

- Code output:

```
S
e
a
o
t
t
```

Loops can be considered as a block of code with an instruction sequence which gets repeated till some defined condition is reached. Loops can help in saving time, reducing errors and making code more readable.

While Loop

While loop runs through the loop till the condition is true.

Example: A simple program to find the even numbers between 0 to 100 in a single line.

```csharp
using System;

namespace MyApplication
{
    class Program
    {
        static void Main(string[] args)
        {
            int i = 1;
            while (i < 50)
            {
                if(i%2==0)
                {
                Console.Write(" " + i);
                }
                i++;
            }
        }
    }
}
```

In this code, variable i is assigned with the value of 1. i value gets incremented by 1 each and every time when the programs runs each and every time which it runs the while loop till variable i value becomes 50. Under the while loop body, i values which are divided by 2 (even numbers) get printed till the value i reaches 50.

Output is as follows:

```
2 4 6 8 10 12 14 16 18 20 22 24 26 28 30 32 34 36 38 40 42 44 46 48
```

Do / While Loop

Do while loop is a variant of While loop where the loop runs once without considering the condition and then runs till the condition is true.

Given below is an example where do/while loop is used. Sample code is as follows:

```csharp
using System;

namespace MyApplication
{
    class Program
    {
        static void Main(string[] args)
        {
            int i = 0;
            do
            {
                Console.WriteLine(i);
                i++;
            }
            while (i < 5);
        }
    }
}
```

This is a simple code where we initialize the integer variable i as 0. Under do block i is written to be printed without checking any conditions. Then i value is incremented by 1. Then it goes to the while loop body where i is checked whether it is less than 5. As soon as i value exceeds 5 it gets terminated from the loop and i will not be printed anymore.

The output is as follows:

```
0
1
2
3
4
```

For Loop

A simple code example where it uses "for" loop is given below.

```
using System;

namespace MyApplication
{
    class Program
    {
        static void Main(string[] args)
        {
            for (int i = 0; i < 5; i++)
            {
                Console.Write(" " + i);
            }
        }
    }
}
```

In this code integer i variable is initialized as 0. Then i is incremented by 1 till i reaches i=5. Till that condition where i < 5 is satisfied, i gets printed on the same line.

The output is as follows:

```
0 1 2 3 4
```

Let's do another example. Let's consider that you need to print the even numbers between 0 to 20. We can use a simple for loop to write that. The sample code is as follows:

```
using System;

namespace MyApplication
{
    class Program
    {
        static void Main(string[] args)
        {
            for (int i = 0; i < 20; i=i+2)
            {
                Console.Write(" " + i);
            }
        }
    }
}
```

In this code, integer i variable is initialized as 0. As the condition i gets incremented by 2 once every time it passes through the loop. Under the for loop there is a condition to check whether i is less than 20. So, till that condition is satisfied i is printed as in the loop body.

The output is as follows:

```
0 2 4 6 8 10 12 14 16 18
```

As another example let's consider that you need to print the numbers in a descending order from 10 to 1. For that we can write a simple code using for loop as follows:

```
using System;
using System.Collections.Generic;

public class Program
{
    public static void Main()
    {
        for(int i = 10; i > 0; i--)
        {
            Console.WriteLine("Value of i: {0}", i);
        }
    }
}
```

Here variable **i** is assigned with value 10 and it gets decremented by 1 each time **(i--)**. And the values are printed making the output as follows:

```
Value of i: 10
Value of i: 9
Value of i: 8
Value of i: 7
Value of i: 6
Value of i: 5
Value of i: 4
Value of i: 3
Value of i: 2
Value of i: 1
```

Exercises to Practice

1. Make a **while loop**. Choose any number between 1 and 10 and set **int a** as your number (ex: int a = 5). Then, with the while loop, increase the number **a** to 100, and print its value to check if it's okay.

2. Make a long string with at least 20 characters (maybe think of your favorite quote and write it down). Then, with a **for loop**, go through each character and find the letter "a". If you find the letter "a", print it out. If there's no letter "a" in the string, print out "b" as the output.

Arrays

Understanding Variables and Arrays

So far, you've seen this type of code **int a = 10** many times now, which is a way of declaring a **variable**. The **int** describes that the number is an integer (i.e. a whole number), the letter **a** is its name, and **10** is the value we're assigning it. In C#, there are different types of **variables,** including:

- **int:** integers - whole numbers without decimals, can be positive or negative
- **double:** numbers with decimals, like 20.55, or 3.14
- **char:** characters, usually containing one letter or number, like "g" or "w"
- **string:** strings store text, like "Hello there!"

With how far C# has come, it now has many different variables thanks to all of its frameworks, though the 4 we've mentioned above are enough for a beginner.

Variables are quite useful. A typical example of their usage would be storing grades, writing down names, birth dates, etc.

However, you can't write more than one value in a single variable. If you wanted to store all of your grades, for example, you'd have to manually declare every variable:

```
int historyGrade = 7;
int geographyGrade = 10;
int englishGrade = 8;
```

What if there was a way to store multiple values of the same type in a single variable? This is where **arrays** come in.

Examples of Array Usage

Arrays allow you to store multiple values in a single variable - and they're really simple to understand too. These are some simple examples of arrays:

```
int[] grades = {7, 10, 8};

string[] favoriteNames = {"Barbara", "Laura",
"Ivy", "Annah", "Eva"};

string[] bestCars = {"BMW", "Porsche", "Mercedes",
"Audi"};
```

With **arrays,** you can store many values in a single variable, which will be incredibly useful when you will be programming bigger projects.

Your programs will become much more powerful if you combine **arrays** with **loops,** which we already learned about in the previous chapter. Let's imagine a situation where you'd like to print out all of your favorite names from the array we wrote above:

```
string[] favoriteNames = {"Barbara", "Laura",
"Ivy", "Annah", "Eva"};
for (int i = 0; i < favoriteNames.Length; i++) {
    Console.WriteLine(favoriteNames[i]);  // will
print the i-th
                                          // name
of the array
}
```

- The code above will print the names in the same order

  we've written them → Barbara, Laura, Ivy···

- Code output:

```
Barbara
Laura
Ivy
Annah
Eva
```

If you've noticed something unfamiliar, it might be the
favoriteNames.Length part of the code. This is just a simple line
of code that accesses the number of elements we have in the
array (in this case it's 5).

Same type of different elements can be stored in one unit by using an array. To declare an array, it is required to define the variable type with square brackets.

string[] fruits;

We now have declared an array in the type of string. Here we can store a large number of values of strings with comma separated.

string[] countries = {"USA", "UK", "India", "Japan" }

In the similar way, arrays can be used to store set of different integers, double, chars etc. in one variable.

int[] myNum = {2, 4, 8, 10, 12}

char[] letters = {'a', 'b', 'c', 'd', 'e'}

double[] values = {10.5, 12.56, 11.78, 15.9}

There is a specific way of accessing the elements in an array. Following sample code shows an example of accessing the elements of an array.

```
using System;
namespace MyApplication
{
  class Program
  {
    static void Main(string[] args)
    {
      string[] fruits = {"Mango", "Banana", "Apple", "Grapes" };
      Console.WriteLine(fruits[0]);
    }
  }
}
```

Here the first element in the first array is named as fruits[0] and the second element as fruits[1] and incremented accordingly. In the sample code fruits[0] access the 1st element of the fruits array. So the output of this simple code is as follows:

```
Mango
```

Array elements can also be changed by simply accessing the elements using the index number. Example code is given below.

```csharp
using System;

namespace MyApplication
{
    class Program
    {
        static void Main(string[] args)
        {
            string[] fruits = {"Mango", "Banana", "Apple", "Grapes"};
            fruits[0] = "Oranges";
            Console.WriteLine(fruits[0]);
        }
    }
}
```

Here the first element of the array which is "Mango" (fruits [0]) is replaced with the "Oranges" and then the new element is accessed and get printed.

The output is as follows:

```
Oranges
```

Arrays have a separate special property to find the number of elements inside an array. Given below is a sample code to illustrate that property.

```
using System;

namespace MyApplication
{
  class Program
  {
    static void Main(string[] args)
    {
      string[] fruits = {"Mango", "Banana", "Apple", "Grapes"};
      Console.WriteLine(fruits.Length);
    }
  }
}
```

Length of an array is taken by *"ArrayName"*. Length function. In the example, array name is "fruits". By fruits.length function we can get the number of elements in the fruits array which is 4. So, the output is as follows.

```
4
```

There is a special function called **foreach** loop, which is used exclusively to loop through elements in an array:

```
using System;

namespace MyApplication
{
  class Program
  {
    static void Main(string[] args)
    {
      string[] fruits = {"Mango", "Banana", "Apple", "Grapes"};
      foreach (string i in fruits)
      {
        Console.WriteLine(i);
      }
    }
  }
}
```

The example above can be read like this: **for each** string element (called **i** - as in index) in **fruits**, print out the value of **i**. So the output will be as follows:

```
Mango
Banana
Apple
Grapes
```

Exercises to Practice

1. Create a basic C# program and run it. Then, make an array of at least 10 elements and print each one of them (you can use loops if you think they'll help you). Change the 3rd element of the array to a different value (if it's 5, make it 7, for example). Print out the entire array.

2. Make an array with your favorite 3 names. Then, check if the 2nd name is shorter in length than the 1st one (hint: use an if statement - nested if needed). Then, do the same for the 2nd and 3rd name. The 1st name should be the shortest, and the 3rd one the longest. The 2nd one should be in the middle.

Strings

In this book so far, you've encountered strings a lot. Now, let's dive deeper into what they are and how they work. In the simplest terms possible, strings are just **variables** used for storing texts - i.e. anything that's more than one character (for single characters, we use **chars**).

It's worth mentioning that strings have their own sets of **methods** that you must use if you intend to operate with them as data. Now that we've got the basics of the theory behind us, let's take a look at some practical examples.

Examples of Strings

Basic C# String Usage

Let's take the simplest usage of a string as our first example:

```
string txt = "abcdefg";
Console.WriteLine("Our string is " + txt);
```

Output:

```
Our string is abcdefg
```

Using String Methods in C#

Strings have many methods that are used in conjunction with them, like **Length, ToUpper(), ToLower(), Concat()**, etc. Take a look at the following example:

```
string firstName = "Mark ";
string lastName = "Donovan";
string name = string.Concat(firstName, lastName);
Console.WriteLine("The name is " + name + " and the
length is:" + name.Length);
```

Output:

```
The name is Mark Donovan and the length is:12
```

Comparing Strings Together

With the method **Equals()**, you can compare the content of two strings and check whether or not they are equal. The **Equals()** returns true or false - true if the strings are identical, false otherwise. Let's take a look at an example:

```csharp
string string1 = "I love C#";
string string2 = "I love C#";
string string3 = "I love C";

Boolean result1 = string1.Equals(string2);
Console.WriteLine("strings string1 and string2 are
identical: " + result1);

Boolean result2 = string1.Equals(string3);
Console.WriteLine("strings string1 and string3 are
identical: " + result2);
```

The output is quite easy to predict. Can you guess it without seeing below?

```
strings string1 and string2 are identical: True
strings string1 and string3 are identical: False
```

Using String Interpolation

If you decide to take a career in programming, you'll probably hear the term interpolation more often. In C#, **string interpolation** refers to using & inserting variables inside of a string. However, to insert a variable inside a string, you must use special symbols, which we will show below:

```
string firstName = "Philipp";
string message = $"Hello {firstName}, welcome"; // this is string interpolation
Console.WriteLine(message);
```

The output is the following:

```
Hello Philipp, welcome home
```

Example On Using Special Characters in Strings

You've probably noticed already that strings are strictly defined with double quotes (this symbol " "). We need one double quote (") to mark the beginning of the string, and another one to mark the end of the string ("). So, what happens if we have a string with a double quote inside of it?

```
string example = "This is an example of a double
quote " usage";
```

If you try compiling the code above, you will get an error. You can even see by the colors that the compiler automatically detects that something is wrong. So, how can we use a double quote inside a string? With an **escape character**.

In C#, the character is the backslash symbol \. When you use it, it tells the compiler to ignore the next char and not compile it. Now, let's run the same code again with the **WriteLine()** function:

```
string example = "This is an example of a double
quote \" usage";
```

Output:

```
This is an example of a double quote " usage
```

Exercises to Practice

1. Think of a sentence you said to a friend or family member today. Write it down in your C# project and print it out. Then, print out its length and see if it's longer than 10 characters.

2. Think of two sentences you said today and write them down as separate strings - **string1** and **string2**. Then, convert all of the characters in the first string to lower letters (hint: see above in the examples) and the second one's characters to upper letters.

3. Write a string with more than one special character inside. Consider using some of the following: *, ", ', ^. If you get an error in your compilation, use **escape characters** to run your code successfully.

4. Create the following string: "Hi there {}, how are you today {}?". Then, create a new string with your favorite name. Use **string interpolation** to fill out the spaces where there are empty brackets → {}. (hint: look at the example for **string interpolation**)

5. Create two strings. One of them should be "Hello there, " and the other one should be "how are you today?". Use the string methods we talked about above to concatenate

them, and create a new string that's composed of both them. Then, print it out.

Math Functions

Programming, as a way to work with a computer, wouldn't exist today without the help of math and mathematicians. So, it's completely expected that the same is true vice-versa, as programming is now used as one of the best tools to improve the field of math as a whole.

This is why you will learn about math functions in C#. Math functions in C# are components of the **Math** class (if this is unclear, reread chapter **classes)**. As we said, a class consists of properties and methods, which we will be using right now in our math functions lesson. Let's go over the basics first.

So, how can you start using the **Math** class? All you have to do is write one line of code - **using System.** The **using System** keyword refers to the namespace *System* where the **Math** class is. Before going through some complex examples, let's take a look at a simple one where you can see how you can start using the **Math** class:

```
using System;
Console.WriteLine(Math.Max(5, 10));
```

Output:

```
10
```

To start using the math class, all we had to write down is **using System**. Then, we used the method Math.Max() which returns the bigger value of the two we give it (using Math.Min() does the opposite). Now, let's dive into some slightly harder examples

Using the *Floor* and the *Square Root* Functions

Almost all programming languages allow for the usage of the **Floor** and **Sqrt** functions. The floor function takes a decimal number (let's say 12.5) and returns its *floor,* i.e. how the number would look without the decimal (in this case, 12).

Square root, on the other hand, is self explanatory, though the function has proven to be *incredibly useful*, simply because there hasn't been found a way to consistently calculate square roots in an accurate manner. Let's take a look at an example:

```
using System;
Console.WriteLine(Math.Floor(500.800));
Console.WriteLine(Math.Sqrt(289));
```

Guessing the output for the first function is pretty easy - but can you do it with the second one? Most people don't know square roots off the top of their heads, but now, there's really no need to know them at all. With the Sqrt() function, we can calculate the square root of the number given to us and return the whole value.

Output:

```
500
17
```

Calculating Cube Roots With C#

Depending on the number of math classes you've taken until now, you might or might not be quite familiar with cube roots. If a square root shows what the number we would receive from multiplying the same number **twice,** a cube root shows the number we'll get if we multiply it **three times.**

As you can already guess, cube roots are a lot harder to calculate than square roots, as they add an entirely new dimension to the picture. Let's take a look at some code:

```
using System;
Console.WriteLine(Math.Cbrt(1000));
Console.WriteLine(Math.Cbrt(4913));
Console.WriteLine(Math.Cbrt(1000000));
```

Can you guess the output? Whether or not you have studied cube roots, it's still quite complicated to do so. No worries though - you've got C# to calculate for you:

```
10
17
100
```

Going Over Other Math Functions

C# has other functions in the **Math** class too, but there's no need to practically show them as most of them are easy to understand.

- Math.Abs() - returns the absolute value of a number

```
Console.WriteLine(Math.Abs(-5));
```

Output:

```
5
```

- Math.Pow() - raises a number to a specific power and returns the value

```
Console.WriteLine(Math.Pow(5, 3));
```

Output:

```
125
```

- Math.Ceiling() - opposite of Math.Floor(), raises the value of the decimal to an **int**

```
Console.WriteLine(Math.Ceiling(5.5));
```

Output:

```
6
```

- Math.Sin() - returns the sine of a number (important for geometry)

```
Console.WriteLine(Math.Sin(90));
```

Output:

```
0,8939966636005579
```

- Math.Cos() - returns the cosine of a number (also important for geometry)

```
Console.WriteLine(Math.Cos(180));
```

Output:

```
-0,5984600690578582
```

Exercises to Practice

1. Write down two integer values and a float - a decimal number. Consider making all of these values double-digit. Regarding the float, use the Math.Ceiling() and Math.Floor() functions and try to calculate the output. Then, calculate the square roots of the integers and try to predict the output.

2. Think of a whole number and write it down as an integer. Then, write the same number in another integer, but with a minus in front of it - i.e. make it negative. Use one of the functions listed above to make the two numbers **equal** (i.e. think of a way to flip the minus to a plus).

3. Use the geometrical mathematical functions we mentioned above (Cos() and Sin()). Experiment with values and try to get them to return the same value.

4. Think of a number from 1 to 10. Then multiply it on paper 3 times over (ex: 5*5*5). Using the Math.Pow() function, check if your answer is correct

5. Write down the number from exercise 4. Then, go through the functions we described above and think of a way to reverse the number and get your original input (if it was 5*5*5, use a function that will return 5 when you input 125).

Pointers

We've got all of the easy subjects out of the way, and now we'll dive right into pointers, which will probably be slightly more challenging than any of the topics we've covered so far. So, let's get some basic theory out of the way first.

Pointers are variables (just like strings, chars, and ints) that hold the **memory address** of a variable. A memory address is the specific place in your computer's memory where the variable in question is stored. So, with pointers, you'll have direct access to your computer's hardware. How cool is that? Let's take a look at some examples.

Examples of Pointers

The following example will consist of everything there is to know about pointers, so make sure you pay close attention:

```
int random = 1234567890;
int* pointer = &random;
Console.WriteLine(pointer); //Shows the address
(where it's stored)
Console.WriteLine(*pointer); //Shows the value of
the address.
```

This is our output:

```
-447224820
1234567890
```

Keep in mind that if you test the same code in your own C# project, you won't get the same result for the first line. Why is that? This is the case because the computer this project is run on is not the same as your computer - therefore, the memory is different also.

The same code might print out at an output of -382213992 (a totally different address) even though you're storing the **exact same value**. This is the power of pointers - being able to directly

access the memory address inside of the computer's hardware.
Let's take a look at another example:

```
int a = 500;
int b = 1000;
int* pointer1 = &a;
int* pointer2 = &b;
Console.WriteLine((int)pointer1);
Console.WriteLine((int)pointer2);
Console.WriteLine(*pointer1);
Console.WriteLine(*pointer2);
```

Output:

```
1815602604
1815602600
500
1000
```

Similarly, in this example, the first two values of the output (1815602604 and 18150602600) refer to the physical address of the values **a** and **b**. The last two values are just the values of the variables **a** and **b** (notice that we can print out the value of a variable through a pointer).

Given below is another simple example of declaring a pointer variable.

```csharp
using System;

namespace MyApplication
{
    class Program
    {
        static void Main(string[] args)
        {
            unsafe{
                int x = 100;
                int* ptr = &x;
                Console.WriteLine((int)ptr);
                Console.WriteLine(*ptr);

            }
        }
    }
}
```

Here variable x is assigned with 100. A pointer variable (ptr) is declared to get the address of the memory location of x variable. Pointer variable is declared by the following code line.

int* ptr = &x;   here & is used to get the memory address of x and then it is assigned to the pointer ptr.

Then by printing (int)ptr command, we can get the memory address of the x variable. And by *ptr command we can get the value of the memory location pointed here.

The output is as follows:

```
1363370640
100
```

Given below is another example where pointers are used.

```csharp
using System;
using System.Collections.Generic;

public class Program
{
    static unsafe void Main(string[] args) {
      int val = 50;
      int* x = &val;

      Console.WriteLine("Data: {0} ", val);
      Console.WriteLine("Address: {0}", (int)x);
      Console.ReadKey();
      }
}
```

Here **unsafe keyword** is added to the main method. Here **val** variable is declared with the value of 50 and integer type x variable is used to get the address of the x variable. Output of the above code can be obtained as follows:

```
Data: 50
Address: 1350713136
```

Given below is an example where two pointers are used to store memory addresses of two variables.

```
using System;
using System.Collections.Generic;

public class Program
{
    static unsafe void Main(string[] args) {
        int m = 10, n=20;
        int* ptr1 = &m, ptr2 = &n;

        Console.WriteLine("Data:    " +m+ ", " + n);
        Console.WriteLine("Address: {0}", (int)ptr1 + ", " + (int)ptr2);
        Console.ReadKey();
    }
}
```

Variable m and n integers are initialized with 10 and 20 respectively. ptr1 and ptr2 is used to store the memory address of m and n. By referring to the code you can see the way of initializing two pointers in one line. Then the data and addresses are printed separately.

The output is as follows:

```
Data:    10, 20
Address: -1937138880, -1937138872
```

Pointers are basically use for saving the memory and it helps in faster execution time since data are manipulated with the addresses and gets the direct access to memory locations.

Exercises to Practice

1. Write down three variables as integers (use any number you like). First, print out the values of their physical address (take a look at example 1), and then print out their original values **using pointers** (take a look at example 2).
2. Write down a variable and print out its address using a pointer. Then, create another variable and do the same. Now, swap the two variables' **physical addresses.** (hint: use **int* pointer1 = &variable2** and vice-versa).

Date and Time

C# programming language is having a **DateTime** structure to work with date and time. The syntax of creating DateTime object is as follows:

```
DateTime dt = new DateTime();//sets a default time of
01/01/0001 00:00:00
```

By using different constructors, we can set the **DateTime** struct using a **DateTime** object that initializes the values. Given below are some examples of different constructors use for setting the date and time.

```csharp
using System;

public class Program
{
    public static void Main()
    {
        //assigns default value 01/01/0001 00:00:00
        DateTime dt1 = new DateTime();
        Console.WriteLine(dt1);

        //assign year, month, day
        DateTime dt2 = new DateTime(2015, 12, 31);
        Console.WriteLine(dt2);

        //assign year, month, day, hour, min, seconds
        DateTime dt3 = new DateTime(2015, 12, 31, 5, 10, 20);
        Console.WriteLine(dt3);

        //assign year, month, day, hour, min, seconds, UTC timezone
        DateTime dt4 = new DateTime(2015, 12, 31, 5, 10, 20, DateTimeKind.Utc);
        Console.WriteLine(dt4);
    }
}
```

In the first constructor date and time is set and the default value gets displayed.

In the second constructor only the date is specified, but the time is not. So, the default value will be printed for the time.

In the other constructors, both the time and date is specified and the date and time will be displayed accordingly.

So the code output will be as follows:

```
1/1/0001 12:00:00 AM
12/31/2015 12:00:00 AM
12/31/2015 5:10:20 AM
12/31/2015 5:10:20 AM
```

There is a special case where if there is a valid string for date and time, we can change the string to a DateTime object. Given below is a sample code for that.

```csharp
using System;

public class Program
{
    public static void Main()
    {
        var str = "10/10/2022";
        DateTime dt;

        var isValidDate = DateTime.TryParse(str, out dt);
        if (isValidDate)
            Console.WriteLine(dt);
        else
            Console.WriteLine($"{str} is not a valid date string");
    }
}
```

Here, we use TryParse() and TryParseExact() methods to see if the string is a valid expression.

isValidDate is a boolean that returns true if the string represents a valid date.

So given below is the output of the above sample code.

```
10/10/2022 12:00:00 AM
```

Classes

Once you've understood everything else there is about C# and you've come to classes, you're well on your way to become a great coder. Now, pay special attention to this subject, as classes are an **absolute necessity** for becoming a good programmer.

Think of classes as "prototypes". We've covered integers, strings, booleans, and much more. What if you could create your own data type? This is exactly what classes allow you to do.

Classes allow you to create new instances of variables (called **objects**) that follow the exact blueprint you defined. A class consists of **properties** (variables) and **methods** (functions inside of the class). There isn't much to learn about classes theoretically, but there is much, much to practice about them.

Examples of Classes

Creating a New Class and Constructing Objects
Let's say you have a bike that you want to somehow define programmatically. This is what classes allow you to do. Take a look at the following example.

```
public class Bike {
    String color;
    int size;
    int age;
    int topSpeed;
}
```

Now, to **construct** an instance of a class (i.e. an object) you need a **constructor**. Let's see how they're made:

```
public Bike(String color, int size, int age, int
topSpeed) {
    this.color = color;
    this.size = size;
    this.age = age;
    this.topSpeed = topSpeed;
}
```

In this constructor, the keyword **this** refers to the current object we're creating. Then, we're manually assigning values to the new object. Now, how can you use this constructor to create a new object?

```
Bike myDadsBike = new Bike("Green", 10, 12, 14);
```

So, we made a new bike called "myDadsBike" that has a **Green color**, is of **size 10,** and is **12 years old.** The last value (14) refers to the top speed this bike can reach in mp/h. If you don't know some of these values, you can set them equal to 0.

Accessing An Object's Values and Printing Them

So far, you've probably understood that classes allow you to create your own prototype for a variable that's totally different from anything we've encountered so far. Why work manually with a variable called bikeColor, bikeSpeed, or bikeSize, when you can put them together in an object called myBike and have them in one place.

Now, let's try to access an object's value and print it:

```
Console.WriteLine(myDadsBike.color);
```

If you try compiling this code, you will get an error saying that you've violated the class's access-level. What does this mean? This means that you cannot access a class or its objects **unless** you've got the permission to do so. So, how can we get the permission to access a variable inside of the class?

- It's simple → just write the keyword **public** right before the "String color" code.

This would look something like this:

```
public class Bike
{
        public String color;
        int size;
        int age;
        int topSpeed;
}
```

Now, we will get the following output:

```
Green
```

Now, let's look at another simple example

```
class Vehicle
{
  string color = "blue";}
```

In this example, Vehicle **class** is created where it can create different objects in the same class. And vehicle has different features like color, weight, model etc. where these are named as **attributes.** Also, car has different functions like driving, parking, braking etc. where they are named as **methods** in C# programming language. Given below is an example where vehicle class can create an object to access the data from the class.

```
using System;

namespace MyApplication
{
    class Vehicle
    {
        string color = "blue";

        static void Main(string[] args)
        {
            Vehicle myObj = new Vehicle();
            Console.WriteLine(myObj.color);
        }
    }
}
```

In this example, Vehicle class is created and a new object called "myObj" is created to get the data from the class. Here dot syntax (.) is used with the object to access the variables/fields inside a class. Also, there can be many objects created for the same class to access the details from the class. Here myObj.color is used to access the color of the vehicle and will be printed as the output.

```
blue
```

Constructor

A special kind of a method known as "Constructor" is used to initialize the objects. With the defining of the constructor, we get the opportunity to initialize the object when it is created.

```csharp
using System;

namespace MyApplication
{
    // Create a Fruits class
    class Fruit
    {
        public string item;  // Create a field

        // Create a class constructor for the Fruit class
        public Fruit()
        {
            item = "Mango"; // Set the initial value for model
        }

        static void Main(string[] args)
        {
            Fruit myObj = new Fruit();  // Create an object of the Car Class (this will call t
            Console.WriteLine(myObj.item);  // Print the value of model
        }
    }
}
```

In the example, Fruit class is created, and a **string variable** called item is created. Since the constructor is a special method with the same name of the class, a constructor is created here with the name of class name "Fruits". Inside the constructor, item variable is initiated with "Mango". When the main method is called by creating an object, the constructor also gets called automatically and sets the initial value. Code output is as follows.

```
Mango
```

Refer the below examples and identify the advantage of using a constructor in initializing the variables.

```csharp
using System;

namespace MyApplication
{
  class Car
  {
    public string model;
    public string color;
    public int year;
    static void Main(string[] args)
    {
      Car Nissan = new Car();
      Nissan.model = "Leaf";
      Nissan.color = "White";
      Nissan.year = 2010;

      Car Toyota = new Car();
      Toyota.model = "Camry";
      Toyota.color = "Black";
      Toyota.year = 2005;

      Console.WriteLine(Nissan.model);
      Console.WriteLine(Toyota.model);
    }
  }
}
```

```
using System;

namespace MyApplication
{
    class Car
    {
        public string model;
        public string color;
        public int year;

        public Car(string modelName, string colorName, int modelYear){

            model = modelName;
            color = colorName;
            year = modelYear;
        }

        static void Main(string[] args)
        {
            Car Nissan = new Car("Leaf", "White", 2010);
            Car Toyota = new Car("Camry", "Black", 2005);

            Console.WriteLine(Nissan.model);
            Console.WriteLine(Toyota.model);
        }
    }
}
```

Both these two codes give the same output as follows:

```
Leaf
Camry
```

Here you can see that without using the constructors, when creating a new object all the variables should be initialized and it makes a lengthy code. When having a constructor, we can easily initialize the variables of the objects as required and make the code look less complicated and neat.

Special keywords known as "Access Modifiers" are used in object-oriented programming languages that set accessibility of classes, methods and other members. Access modifiers are specially used syntax which facilitate the encapsulation of components.

Modifier	Description
public	Code is allowed to be access for all classes
private	Code is only accessible in the same class
protected	Code is accessible within the same class or a class inherited from the same class
internal	Limits the access of the code within the current project assembly

Most of the object-oriented programming languages have four main concepts: **Encapsulation, Polymorphism, Inheritance** and **Abstraction.**

Encapsulation

Encapsulation is a concept used in Object-oriented programming in order to protect the variables, methods and other data in a class from hiding the data or not allowing users to change the details of the code. This can be simply considered as a capsule which covers the drug inside it. By referring to the below example, let's try to understand this encapsulation concept.

```
using System;

namespace MyApplication
{
  class Vehicle
  {
    private string model;  // field
    public string Model    // property
    {
      get { return model; }
      set { model = value; }
    }
  }
  class Program
  {
    static void Main(string[] args)
    {
      Vehicle myObj = new Vehicle();
      myObj.Model = "Suzuki Estilo";
      Console.WriteLine(myObj.Model);
    }
  }
}
```

In the example, a string variable "model" is created with the access modifier of **private** keyword. So, this variable cannot be accessed by any other class to change the value. So, for initializing and assigning the values for private variable, special methods named as **get** and **set** are used. Here, **model** private variable value is set and get by a method called **Model** (which is the same name as variable). Here **get** method is used to return the value of model variable. **Set** method is used to assign a value to the variable **model**. In the program class in the example, an object named as **'myObj'** is created in **Vehicle class.** Then using the **Model method**, variable **model** is set with the value of

"Suzuki Estilo" which helps to access and change the private variable in the **Vehicle class**. Then, the output is as follows:

```
Suzuki Estilo
```

Inheritance

Inheritance is a concept in object-oriented programing where there are **derived classes (child classes)** which are inherited with the attributes, methods from a **base class (parent class).** This concept helps to reduce the replicate of the codes again and again in different classes. Given below is an example to get a clear idea about this concept.

```csharp
using System;

namespace MyApplication
{
  class Vehicle  // Base class
  {
    public string brand = "Toyota";  // Vehicle field
    public void drive()              // Vehicle method
    {
      Console.WriteLine("I can drive fast!");
    }
  }
  class Car : Vehicle  // Derived class
  {
    public string modelName = "Lexus";  // Car field
  }
  class Program
  {
    static void Main(string[] args)
    {
      // Create a myCar object
      Car myCar = new Car();

      // Call the drive() method (From the Vehicle class) on the myCar object
      myCar.drive();

      // Display the value of the brand field (from the Vehicle class) and the value of the modelName from the Car class
      Console.WriteLine(myCar.brand + " " + myCar.modelName);
    }
  }
}
```

In this example, **Vehicle** is the **Base Class** and there are different attributes like **Brand** which is initialized as "Toyota" and a method with the name of **drive**. There is another derived class named **Car** which is having a different variable named **modelName**. Since Car class is a derived class of Vehicle, it can access all the attributes and methods in the Vehicle class as well. So in the main method, an object is created and call the drive method and value of brand field from Vehicle class. So the output is as follows:

```
I can drive fast!
Toyota Lexus
```

Inheritance concept is mainly used in the code where it is required to have code re-usability. In the place where same code is repeated in different classes, we can get the same properties inherited from the base class hence reducing the repeatability of the code.

Polymorphism

Polymorphism is also one of the main concepts in object-oriented programming where it considered about something which occurs in several different forms. Polymorphism occurs where many classes are related to each other by inheritance.
Given below is an example for an instance where we use polymorphism.

```csharp
using System;
namespace MyApplication
{
    class Animal  // Base class (parent)
    {
        public void animalMove()
        {
            Console.WriteLine("The animal can move");
        }
    }

    class Dog : Animal  // Derived class (child)
    {
        public void animalMove()
        {
            Console.WriteLine("The dog moves by walking");
        }
    }

    class Fish : Animal  // Derived class (child)
    {
        public void animalMove()
        {
            Console.WriteLine("The fish move by swimming");
        }
    }

    class Program
    {
        static void Main(string[] args)
        {
            Animal myAnimal = new Animal();  // Create a Animal object
            Animal myDog = new Dog();  // Create a Dog object
            Animal myFish = new Fish();  // Create a Fish object

            myAnimal.animalMove();
            myDog.animalMove();
            myFish.animalMove();
        }
    }
}
```

In this example, there are three classes called Animal which is the base class and two derived class called Dog and Fish. Here there are same method named as "animalMove" in all three classes. And an object is created from each class and animalMoove method is called. The following output is received.

```
The animal can move
The animal can move
The animal can move
```

After referring to the output we can say that it is not what is expected by creating three objects for different classes. But it happening for a reason. Since all the methods are having the same name from the base class, every method with the same name in derived classes were replaced by the base class method. That is the reason for getting this output mentioned above.

Let's refer the below example and analyze the output.

```csharp
using System;

namespace MyApplication
{
    class Animal  // Base class (parent)
    {
        public virtual void animalMove()
        {
            Console.WriteLine("The animal can move");
        }
    }

    class Dog : Animal  // Derived class (child)
    {
        public override void animalMove()
        {
            Console.WriteLine("The dog moves by walking");
        }
    }

    class Fish : Animal  // Derived class (child)
    {
        public override void animalMove()
        {
            Console.WriteLine("The fish move by swimming");
        }
    }

    class Program
    {
        static void Main(string[] args)
        {
            Animal myAnimal = new Animal();  // Create a Animal object
            Animal myDog = new Dog();  // Create a Dog object
            Animal myFish = new Fish();  // Create a Fish object

            myAnimal.animalMove();
            myDog.animalMove();
            myFish.animalMove();
        }
    }
}
```

This is similar to the previous code. But there are few keywords which were not in the previous code.

- There is a keyword **Virtual** in animalMove method in the **Animal class (base class)**

- There is a keyword **override** in animalMove method in **derived classes**

By using these keywords base class methods will not override the derived class methods as in the previous example. The output will be as follows:

```
The animal can move
The dog moves by walking
The fish move by swimming
```

Abstraction

Abstraction is also a key concept in object-oriented programming where data is hidden from user and only the important data is shown to the user. Here, **Abstract** keyword is used for classes and methods.

- **Abstract class:** This class is a restricted one which does not allow to create the objects. This must be inherited from another class to access this abstract class
- **Abstract method:** This can only be used in an abstract class and it doesn't have a method body and it is provided from the derived class.

Let's consider the below example to understand the use of abstract classes and methods.

```
using System;

namespace MyApplication
{
    // Abstract class
    abstract class Human
    {
        // Abstract method (does not have a body)
        public abstract void humanName();
        // Regular method
        public void age()
        {
            Console.WriteLine("15 years old");
        }
    }

    // Derived class (inherit from Human)
    class Child : Human
    {
        public override void humanName()
        {
            // The body of humanName() is provided here
            Console.WriteLine("Name: John");
        }
    }

    class Program
    {
        static void Main(string[] args)
        {
            Child myChild = new Child();   // Create a Child object
            myChild.humanName();
            myChild.age();
        }
    }
}
```

In this example, it can be seen that there is an **abstract class** called Human (It is also the base class). In Human class there are both regular and abstract methods. Child class which is a derived class from Human class is used to access this **abstract class (Human class).** By using the derived class, we can call the regular method and abstract method in abstract class. The output is as follows:

```
Name: John
15 years old
```

Exercises to Practice

1. Create a class named **Car** that has the following properties inside: color, manufacturer, yearOfManufacturing, engineType (if you're not sure about the details, you can leave them blank). Then, create at least 3 objects with a **constructor** and place the objects in an array (hint: check **lesson arrays**)

2. Create another class called **MobilePhone** and think of at least three properties to put inside. Create a constructor for the class. Then, make at least five objects of the phone class and change some of their properties (hint: use the **public** keyword as in the last example).

Lists

List is a collection of strongly typed objects which can be accessed using the index. Also lists are having different methods for sorting, searching and modifying. It is also considered to be a generic version of ArrayList that comes under the namespace of *System.Collection.Generic.*

Given below is a simple example code of how to create a list consisting with even numbers below 10.

```
using System;
using System.Collections.Generic;

public class Program
{
    public static void Main()
    {
        // adding elements using add() method
        List<int> evenNumbers = new List<int>();
        evenNumbers.Add(2);
        evenNumbers.Add(4);
        evenNumbers.Add(6);
        evenNumbers.Add(8);

        Console.WriteLine("No of elelemts: "+ evenNumbers.Count + " even
numbers");
    }
}
```

In this example, List is initiated with the type of elements that is planned to be stored in it. Since in the example, there we going to store numbers, we have used int type for the List. And by using **Add()** method we can store new elements to the list. And by **Count()** method we can count the number of elements in the List. The output is as follows:

```
No of elelemts: 4 even numbers
```

Let's consider another example of creating a **string type list** to store the main cities in the country.

```csharp
using System;
using System.Collections.Generic;

public class Program
{
    public static void Main()
    {
        // adding elements using add() method
        List<string> cities = new List<string>();
        cities.Add("Colombo");
        cities.Add("Matara");
        cities.Add("Galle");
        cities.Add("Kandy");

        Console.WriteLine("No of elelemts: "+ cities.Count);
    }
}
```

In this example, List is defined for string since the storing elements are names of the cities. Using the Add() method elements are added to the list same as in the previous example. Given below is the output of this simple code:

```
No of elelemts: 4
```

Same as in arrays using index, we can access the elements of the list. Given below is an example for that.

```csharp
using System;
using System.Collections.Generic;

public class Program
{
    public static void Main()
    {
        // adding elements using add() method
        List<string> cities = new List<string>();
        cities.Add("Colombo");
        cities.Add("Matara");
        cities.Add("Galle");
        cities.Add("Kandy");

        Console.WriteLine("No of elelemts: "+ cities.Count);
        Console.WriteLine("Elements of the List are: ");
        Console.WriteLine(cities[0]);
        Console.WriteLine(cities[1]);
        Console.WriteLine(cities[2]);
        Console.WriteLine(cities[3]);
    }
}
```

Here cities[0] represent for the first element of the list and other elements are named as cities[1], cities[2], cities[3] respectively. So the output is as follows:

```
No of elelemts: 4
Elements of the List are:
Colombo
Matara
Galle
Kandy
```

If you need to insert a new element to the list in middle, then you need to use the **Insert()** method as in the following example.

```csharp
using System;
using System.Collections.Generic;

public class Program
{
    public static void Main()
    {
        var fruits = new List<string>(){ "Mango", "Apple", "Banana", "Grapes" };

        fruits.Insert(1, "Oranges");// inserts "Oranges" at 1st index: after Mango.

        foreach (var num in fruits)
            Console.WriteLine(num);
    }
}
```

In this example code, fruits list is created with some string values. And we have a requirement of adding "Oranges" as the second element of the list. For that we use Insert(1,"Oranges") method. Here 1 is the position of the element in the list. Then every element is printed in fruits list will be printed after adding "Oranges".

```
Mango
Oranges
Apple
Banana
Grapes
```

There is a separate method for removing the elements from a list. Given below is an example where **Remove()** methods are used.

```
using System;
using System.Collections.Generic;

public class Program
{
    public static void Main()
    {
        var fruits = new List<string>(){ "Mango", "Apple", "Banana", "Grapes" };

        fruits.Remove("Mango"); // removes Mango element from a list

        fruits.RemoveAt(2); //removes the 3rd element (index starts from 0)


        foreach (var num in fruits)
            Console.WriteLine(num);
    }
}
```

There are two basic methods for removing the elements from a list. They are

- **Remove(*element*):** All the elements with this name will be removed from the list
- **RemoveAt(*number*):** Element with the position as in *number* will be removed from the list

In this example, by **fruits.Remove("Mango")** will remove the Mango from the list and by **fruits.RemoveAt(2)** will remove the 3rd element of the list (new list which is after removing Mango). The 3rd element in the new list is Grapes. So the output is as follows:

```
Apple
Banana
```

Dynamic Variables

Dynamic variables are having the facility of changing the types at run time based on the assigned value. Dynamic variables are normally defined with the **dynamic keyword**. The syntax is as follows:

dynamic myDynamicVar = 10;   *here myDynamicVar is the variable name*

Normally dynamic type escapes at compile time and gets its type resolved at run-time. The following example shows how that the dynamic variable changes at run-time when the values are assigned to those variables.

```
using System;

public class Program
{
    public static void Main()
    {
        dynamic MyDynamicVar = 15;
        Console.WriteLine("Value: {0}, Type: {1}", MyDynamicVar, MyDynamicVar.GetType());

        MyDynamicVar = "Hello World!!";
        Console.WriteLine("Value: {0}, Type: {1}", MyDynamicVar, MyDynamicVar.GetType());

        MyDynamicVar = true;
        Console.WriteLine("Value: {0}, Type: {1}", MyDynamicVar, MyDynamicVar.GetType());

        MyDynamicVar = DateTime.Now;
        Console.WriteLine("Value: {0}, Type: {1}", MyDynamicVar, MyDynamicVar.GetType());
    }
}
```

In the example code, a dynamic variable with the name of **MyDynamicVar** is initialized here. Then different values are assigned to that variable and see how the variable type changes with the assigning value during run-time.

In the first line, variable is assigned with 15 hence it is an integer type. Then "Hello World!!" which is a string. And true value is assigned to MyDynamicVar variable hence making the variable type Boolean. At last dynamic variable is assigned with a DateTime making the variable type as DateTime.

Given below is the output of the code.

```
Value: 100, Type: System.Int32
Value: Hello World!!, Type: System.String
Value: True, Type: System.Boolean
Value: 9/10/2022 7:01:38 PM, Type: System.DateTime
```

Conclusion

We'd like to genuinely thank you for paying attention and reading this book. The team behind it spent hours building & composing it, so we hope that it helped out.

If you went over and understood most of the contents of this book, you now great knowledge in C# and an amazing foundation when it comes to programming in general.

So far in this book, we covered all of the major concepts of programming, including:
- Creating a work environment to start coding,
- The basics of C#,
- Conditionals,
- Variables,
- Arrays,
- Loops,
- Strings,
- Pointers,
- Classes,
- and more

If you enjoyed this book, please don't hesitate to give us feedback. Thank you for your time!

The end... almost!

Reviews are not easy to come by.

As an independent author with a tiny marketing budget, I rely on readers, like you, to leave a short review on Amazon.

Even if it's just a sentence or two!

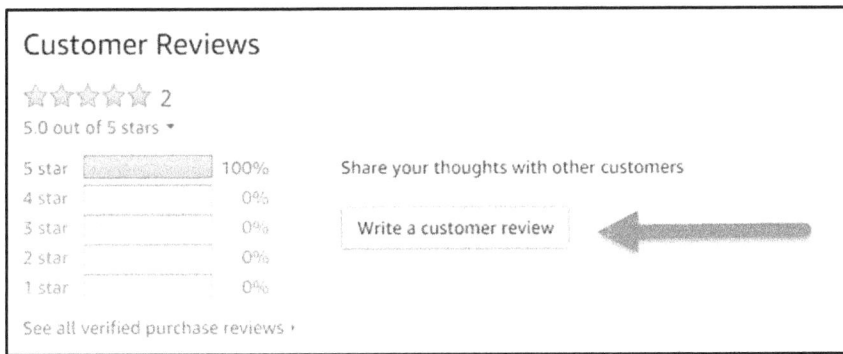

So, if you enjoyed the book, please...

>> Click here to leave a brief review on Amazon.

I am very appreciative for your review as it truly makes a difference.

Thank you from the bottom of my heart for purchasing this book and reading it to the end.

.

www.ingramcontent.com/pod-product-compliance
Lightning Source LLC
Chambersburg PA
CBHW061022220326
41597CB00017BB/2446